My gift to you. Best wishes for
CEO Space Executive Director c

Leslie Knight

Leslie Knight
www.LeslieKnight.com

NAVIGATING THE I.T.
MINEFIELD

Navigating the I.T. Minefield

Leslie Knight

Tendril Press
Aurora, CO

Navigating the I.T. Minefield:
Straight Talk for the Small Enterprise

www.knightpm.com and www.itminefield.com

To order the support CD which is refferenced within this book
please visit *www.itminefield.com*

Published by Tendril Press™
www.TendrilPress.com
PO 441110
Aurora, CO 80044
303.696.9227
Educational & Corporate Quantity Discounts available

First Publishing: 2010
Printed in the USA

ISBN 978-0-9822394-9-0

Cartoons provided by:
Grantland Grant Brownrigg
http://grantland.net/

and

I.T. Doesn't Happen To Me
Jamie Buckley

Contents

Illustrations

Cartoons & Figures

Tables & Charts

Acknowledgements

As with any book, it is rarely, if ever, the effort of just one person. I was blessed along the way with the special insights of several different people. I owe a debt of gratitude to:

Lee Roberts, CEO of MerchantMetrix, Inc. (www.merchantmetrix.com) for quickly launching our website (www.ITMinefield.com) as well as contributions to the chapter on websites. His search engine friendly methods will increase our sales as it has done for other clients. Sheridan Broderick, thank you for making the changes we needed in a timely manner.

Michael Price, CEO of MPA Networks (www.mpa.com) who took the time to review and provide input on Disaster Recovery and Planning prior to leaving for vacation. As the oldest and most successful service provider in Silicon Valley for small enterprises his insights were invaluable. His experience in helping small enterprises recover from disasters was the perfect balance to my experience with larger enterprises.

Jerry Hayward, CEO of Convertabase (www.convertabase.com), Walter Williams, CEO of CHB Consulting (www.chb-consulting.com), and Dean Bickmore, Sales Manager of SilverTree Technology (www.silvertree.net)...your thoughts and examples influenced the content in each chapter.

Scott Degraffenreid, CEO of Necessary Measures, for his creative suggestions which simplified complex topics, developed the theme and kept me on track.

Grant Brownrigg and Jaime Buckley for their on target and humorous illustrations.

Karin Hoffman, CEO and Creative Director, Robin Hoffman (editor) and Kris Harmon (cover design) of A J Images Business Design and Publishing Center, Inc. Home of Tendril Press, LLC publishing house (www.ajimagesinc.com www.tendrilpress.com). This book came together in a fairly short period of time. Thank you, ladies.

Larry and Joan Knight, my parents. Thank you for your love, encouragement and support.

Last, but not least, a special thank you to the Dohrmann's for creating CEO Space, an organization dedicated to the success of the entrepreneur. I met many of the contributors to this project through CEO Space. Without them, this book would not exist.

Many thanks to you all.

Introduction

Does this sound familiar? "I'm ready to start my own business. I need to get some computers and find some help setting things up." Your eager-to-help friend, Joe enthusiastically jumps in, "You should call my brother's son-in-law to help you. It won't be that expensive, and you'll be up and running in no time!"

Low-cost fits right in line with your start-up budget. And fast. Fast is what you want. It sounds like a great deal...and it might be until you have your first problem, and Joe's brother's son-in-law is nowhere to be found.

Or, you sit down with a savvy Information Technology (IT) or computer specialist or web designer, someone who dazzles you with his extensive knowledge yet demonstrates no interest in you or your business needs. You are left completely confused, wondering why you need to spend $20,000 on technology.

Unfortunately, most small business enterprises and non-profit organizations (10 or fewer computers) find themselves sitting in the middle of a technology minefield. Without in-house tech support, they are forced to count on contractors who are not likely to be around very long or find they are at the mercy of high-end service providers who present solutions in impressive, yet bewildering technical jargon. These situations expose the smaller enterprise to many dangers including cost overruns, lack of documentation and disaster planning, and ineffective security.

The purpose of this book is to help you navigate the IT minefield. What you don't know or your provider doesn't know CAN hurt you, and what you don't know that you don't know can be devastating.

How severe the damage and the cost of recovery will be determined by your ability to detect and sidestep the mines. In this book, you will:

- Discover the "blind spots" that expose your business to unnecessary risks.
- Learn how to effectively communicate your business requirements to IT experts.
- Understand security issues and how to protect your systems and proprietary information.
- Develop strategies to plan for and manage system failures so they don't bury your business.

GRANTLAND®
"IT IS VERY IMPORTANT TO ANALYZE RISK AND EVALUATE POTENTIAL PROBLEMS.

"SO TAKE A LOOK AT YOUR COMPANY AND ALL OF THE BAD THINGS THAT MIGHT HAPPEN,

"LIKE FLOODS, FIRE, HURRICANES, TORNADOES, EARTHQUAKES, THEFT, EMBEZZLEMENT, QUALITY PROBLEMS, ACCIDENTS, LIABILITY CLAIMS, PRODUCT RECALLS...."

Chapter 1

Uncovering the Mines

If you needed to fly somewhere, who would you prefer be seated in the cockpit? A great pilot who knows he is great? A mediocre pilot who knows he is mediocre or a mediocre pilot who thinks he is great?

Most likely, you would not pick the latter. Great pilots who know they're great are confident in their true abilities. Mediocre pilots who know they're mediocre are aware of their limitations and operate accordingly.

The dangerous pilot is the mediocre one who thinks he is great. He is blind to his limitations, and his overconfidence could recklessly place everyone on the plane at great risk.

The IT Mine Detector™ is designed to reveal blind spots in your awareness and potential threats that may do real harm to your business.

As you complete the assessment, don't be overly concerned with getting a great "score." No answer is "right" or "wrong". Be honest and pick the answers that best reflect your situation.

The IT Mine Detector™

A copy of the assessment is available on the support CD if you wish to use it instead. Please honor the copyright and only use it for yourself.[1]

1. How important is it to you to spend your technology dollars wisely?
   a. Very important
   b. Somewhat important
   c. Not important

2. How confident are you that you have gotten great value for your technology dollars?
   a. Very confident
   b. Somewhat confident
   c. Not at all confident

3. Do you have a strategic plan for your business?
   a. Yes
   b. No

4. To what extent does your strategic plan guide your technology decisions and purchases?
   a. We consistently make purchases based on the plan
   b. I have not really thought about using the strategic plan for that purpose
   c. It has no bearing

5. Who handles tech support?
   a. We have in-house IT personnel
   b. We have a support contract with a vendor
   c. A friend of a friend

6. How is vendor staffing managed?
   a. We have one primary vendor who manages the other vendors
   b. We coordinate the efforts of several vendors
   c. Support is requested on an as needed basis

7. How important is thorough software, hardware and process documentation?
   a. Very important
   b. Somewhat important
   c. Not Important

8. How confident are you that hardware, software and process documentation are complete?
   a. Very confident
   b. Somewhat confident
   c. Not at all confident

9. Who is responsible for maintaining software, hardware and process documentation?
   a. We are
   b. Our service provider
   c. We have not discussed it

10. Do you and your IT support vendor have a transition plan in place in case the vendor is no longer available?

    a. Yes, we have a plan
    b. No, we don't have a plan
    c. We don't use a support vendor

11. How confident are you that your service provider is looking out for your best interests?

    a. Very confident
    b. Somewhat confident
    c. Not at all confident

12. When you place a service call, do you receive documentation from the provider?

    a. Always
    b. Sometimes
    c. Rarely

13. How important is your ability to recover after a disaster?

    a. Very important
    b. Somewhat important
    c. Not important

14. How confident are you that you could recover critical files if they become lost or damaged?

    a. Very confident
    b. Somewhat confident
    c. Not at all confident

15. How often do you backup your critical files?

    a. Daily or as they change
    b. Regular schedule, but not daily
    c. Infrequently or not at all

16. Where do you store your backup files?

    a. Offsite, away from the office
    b. In a fireproof safe
    c. Loosely in the office

17. How often do you test critical file recovery?

    a. Once a year
    b. Rarely
    c. Never

18. How prepared are you for a disaster?

    a. We've developed and tested our plan.
    b. We have a plan.
    c. We're toast.

19. How important is the security of your network and computers?

    a. Very important
    b. Somewhat important
    c. Not important

20. How confident are you in the security of your network and computers?

    a. Very confident
    b. Somewhat confident
    c. Not at all confident

21. Do you have a defined security policy for your company?

    a. Yes
    b. No
    c. What is a security policy?

22. Do you enforce your security policy?

    a. Yes, all personnel must comply
    b. Sometimes
    c. Rarely

23. How has your policy been communicated to your employees?

    a. Formally and informally
    b. Formally (meetings & written documentation)
    c. Informally (talking around the water cooler)

24. Are your computers locked down to prevent virus infection?

    a. Yes
    b. No
    c. I don't know

25. How important is the website to your business?

    a. It is/will be a primary revenue source
    b. It is primarily a marketing tool
    c. It is a yellow pages ad

26. How confident are you that your website can be recovered after a disaster?

    a. Very confident
    b. Somewhat confident
    c. Not at all confident

27. Is your website achieving the results you expected?

    a. Yes, the data shows it is meeting our goals
    b. No, it is falling short of our expectations
    c. I can't really tell

28. Has the disaster recovery plan for your website ever been tested?

    a. Yes
    b. No
    c. I don't know

29. Do you have a transition plan in case the vendor hosting your site closes business?
    a. We host the site internally
    b. Yes
    c. No

30. Do you have all of the documentation for your site (code, processes, etc.)?
    a. Yes, we host the site
    b. Yes, the vendor has provided it
    c. No

31. How important is keeping operating system and application maintenance up to date?
    a. Very important
    b. Somewhat important
    c. Not important

32. How confident are you that maintenance is current?
    a. Automatic updates are enabled
    b. Updates are scheduled regularly
    c. Not at all confident

33. How old is your operating system?
    a. Current (e.g. Windows® XP/Vista)
    b. One version back (Windows® ME)
    c. Two versions or more back (Windows® 98 or older)

34. How many different operating system versions does your company support?

    a. 1
    b. 2
    c. 3 or more

35. How up-to-date is your most critical application?

    a. Current
    b. One version back
    c. Two or more versions back

36. Who drives the upgrade cycle in your business?

    a. The company does based on the strategic plan
    b. The application vendor
    c. My IT support vendor

37. How important is keeping all of your software legal?

    a. Very important
    b. Somewhat important
    c. Not important

38. How confident are you that all of your software is legal (not pirated)?

    a. Very confident
    b. Some confident
    c. Not at all confident

39. Which response describes software piracy?

    a. Copyright infringement
    b. The unauthorized copy and distribution of software
    c. Reusing my old software on a new computer

40. What are you doing to prevent the installation of pirated software?

    a. Computers are locked down to prevent unauthorized software installation
    b. We trust employees to do the right thing
    c. Nothing special of which I'm aware

41. What are you doing to prevent piracy of purchased software?

    a. All discs are stored in a secured area
    b. Each employee is responsible for protecting the software
    c. No special measures taken yet

42. What are you doing to prevent piracy of your proprietary applications and information?

    a. Special measures are in place to prevent piracy
    b. Access is restricted based on functional job requirements
    c. I didn't realize the need to protect my own assets

Which Mines Did You Uncover?

This assessment tests for three things:

1. Your perception of the importance of a subject to your business.
2. Your confidence in a given area.
3. Specific steps you may have already taken to protect your business.

To score the assessment, record your points in the table. Give yourself 3 points for each "a" response, 2 for each "b" and 1 for each "c" answer. If you start with question 1 and fill in the row with each succeeding response, you will fill in the whole table.

	Importance	Confidence				Total
Cost Management (Questions 1-6)						
Documentation (Questions 7-12)						
Disaster Preparedness (Questions 13-18)						
Security Policy (Questions 19-24)						
Website (Questions 25-30)						
Maintenance (Questions 31-36)						
Legal Compliance (Questions 37-42)						

Table: 1.1— Assessment Score Card

Low values indicate that you are aware that your business is vulnerable, possibly in ways you had not considered. When you identify the blind spots you can take action to improve.

High values generally point to two situations. Either the business has employed an effective IT service provider (either outsourced or on staff) to support the business or the business owners are like the mediocre pilots who think they're great, when they are not. Unfortunately, their lack of awareness and denial is exposing their business to unnecessary perils.

Adding up the values to see your score is not as important as noticing the relationship between the values. How many times did you indicate something was very important or important (a value of 3 or 2), yet your confidence in your company's performance in that area is low (a value of 2 or 1)? High importance and low confidence areas may need immediate attention. If importance is low and confidence is high, you might choose to focus on other areas first.

Use the four columns to the right of the confidence column to rate the strength of the steps you have taken. See if your confidence is justified. Add the four values together and place the sum in the last column. The higher the total, the stronger the measures you have taken to protect your business. A total score of 12 is the maximum in any area. Any score below 12 suggests areas where you might need to do some work.

The Mines

The most-common IT mines fall into eight broad categories.

- Technology cost management.
- Security policy.
- Documentation of IT assets and work processes.
- Engagement in the technology planning process.
- Delegation of technology planning to a service provider.
- IT project management.
- Disaster planning and recovery.
- Legal compliance with software copyrights.

If you have discovered weaknesses in any of these areas through completing the assessment, remember the upside is that you have uncovered the mines that could threaten your business. As you read through this book, you will find specific actions you can take to limit your exposure.

In Chapter 11, *Clearing the Minefield,* you will have an opportunity to quickly take stock of where you are today and list the actions you need to take. As you are reading, you might jot down your thoughts from each section in a separate notebook.

1. IT Service Providers may license the assessment for use with their clients without purchasing the book, or if you wish to purchase the support CD mentioned within please contact us at www.itminefield.com

GRANTLAND®

THERE ARE FOUR BASIC STEPS IN STRATEGIC PLANNING. FIRST, WE GATHER THE FACTS.

SECOND, WE MAKE PROJECTIONS.

THIRD, WE WRITE THE REPORT.

FOURTH, WE LOCK IT UP.

Chapter 2

Am I Being Ripped Off?

Are you being ripped off? The answer is generally, "No." Most often, excessive costs result from a failure to communicate. Communication can be a challenge when IT professionals use jargon and acronyms. Even minor breakdowns can leave the consumer feeling ripped off, so the larger the project, the more crucial it becomes to clearly state your needs and maintain a mutual understanding throughout your IT relationship.

After working in a firm for many years, a friend decided to start his own practice. All he knew was that he needed a couple of computers, so he asked his clients for references. He received several bids to purchase and install the equipment that varied widely in cost. After a few weeks of mulling over the bids and trying to make a decision, he asked me to look and provide some input. We spent some time talking about his current technology and what he was trying to do and crafted a plan that was less expensive than the other bids.

In defense of the bidders, my friend probably went to them and said, "I need two computers and the ability to share files." They probably then attempted to establish more specific parameters, asking questions in mind-numbing technical jargon. Having lost his attention, they proceeded without further input and devised a plan to fit what they thought he needed.

What was different? The approach. Most business consumers do not care about the underlying computing technology except when they have a special or specific need. For any business (even the Fortune 1000), Information Technology exists for only one purpose: to support the business strategy. Think of it as a pyramid. At the top is your business strategy. The business strategy determines the applications and processes you will need to succeed. Some applications you will purchase off the shelf and modify. Others you may build because of their strategic importance.

The applications you employ as well as security and disaster planning will determine the underlying

Information Technology Architecture, what exactly you need in terms of computers, networks, security and other hardware, as well as what you will outsource and what you will support in house.

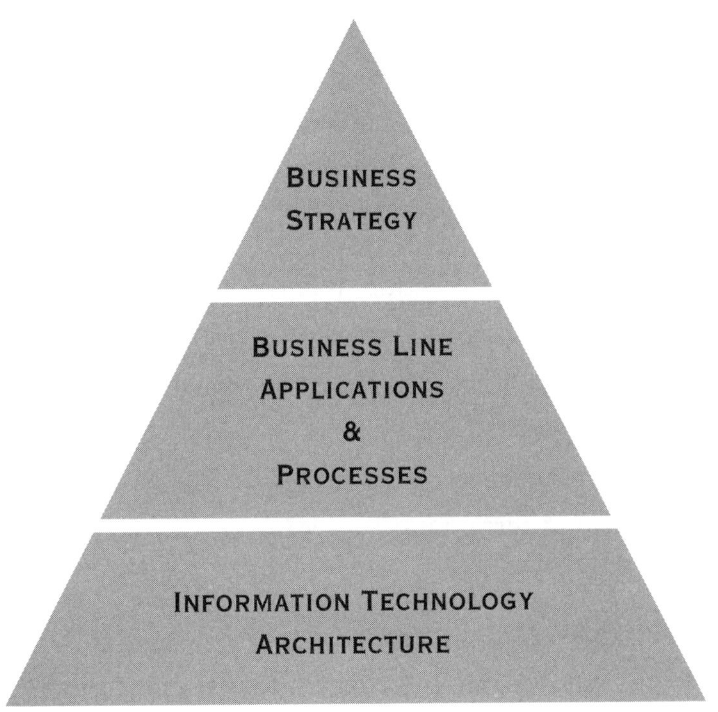

Figure: 2.1— Technology Supports Strategy

That being said I assume that you have a strategic plan for your business.[1] You know who your clients are and how you uniquely fill a role for them. You have a strategy, a plan for winning in the marketplace against your competition. You also have a vision for the growth of your business. Now the question becomes, how can technology support your strategy?

Understanding Your Business

Every business is unique. To adequately meet your needs, a service provider needs to understand your business and the role technology plays. Before you contact a service provider, consider these questions:

- What functions does technology support?
  - Collaboration among your employees?
  - Information exchange between you and your clients?
  - Financial record keeping?
  - Long-term data storage?
  - What else?
- What do your clients expect from you? What is your promise to them? How does technology support delivering on the promise?
  - Confidentiality?
  - Security?
  - Rapid response?
  - What else?

- When do you plan to replace the equipment?
  - Do you plan to expense or depreciate it?
  - Are laptops for sales and remote personnel "disposable" assets? If so, you should only spend what you need to make them functional.
  - Do you expect your needs to change significantly in the next three years? If not, you might spend a little more money on memory, monitor and hard drive to extend the useful life of the computer.

- What do your employees really need in terms of technology to perform their job functions? Are your employees basic users, power users or a mix? This will impact the configuration of your computers.
  - Basic users tend to use the applications just as they come delivered on the computer or as purchased. Their utilization of the computer is limited to email, the internet, word processing and typical business applications.
  - Power users have the skills to build their own applications to increase the usefulness of an application. They also may use resource intensive applications such as CAD/CAM. These users will require more memory and a faster processor than your basic user.

- Do you need to share files, printers and other hardware among employees?
  - If so, you need a server. However, you might be able to convert an older computer to a server with a memory and hard drive upgrade rather than purchasing a computer with a network operating system.
- What are the security requirements for your business and your industry (e.g. HIPAA, SOX, PCI-DSS)?
  - These requirements will influence network choices, security policy, disaster planning and website and application development.
- How long can a critical process, application or file be unavailable before your business is adversely impacted?
  - Your answer will affect decisions concerning disaster planning, environment configuration, Uninterruptable Power Source (UPS) purchases and support contracts.
  - Concerning UPS purchases, one can overspend without gaining additional value or protection. The UPS for a desktop or docked laptop should be able to sustain a computer and monitor for 15 minutes to permit an orderly shutdown. The

UPS for a server should be able to sustain the server and monitor for 30 minutes. Many UPS devices can detect a power outage and initiate an orderly shutdown in your absence.

- If your business cannot tolerate an extended power outage you should investigate alternative power sources, such as diesel generators.

- Do you need to provide remote access for sales personnel or other off-site employees? This will impact decisions concerning network design.

- Do you have any special requirements for high availability (e.g. 24/7) access? This answer affects the entire IT infrastructure.

The better you understand your business, the more clearly you can communicate your requirements to the service provider.

What Else Can I Do?

The first steps to controlling your costs are to understand the needs of your business and allow your business strategy to guide your expenditures. Below are several steps you can take to help you be confident you are controlling costs well.

- When possible, order hardware and replacement parts directly from the manufacturer or computer store. Most service providers place a 5% to 20% premium on ordering your equipment and replacement parts to cover their costs in placing your order for you.

- Compare bids from three or four vendors.
  - If they want to charge you a nominal fee for preparing a bid, pay it. Don't expect them to do the leg work so you can go hire another vendor.

- Avoid vendors who want to charge you by the hour. Most vendors should be able to provide a firm bid based on size and scope of the project.
  - If you don't have a service contract with the vendor, with a monthly support fee, you will pay an hourly rate, usually starting at $150 for a minimum of four hours.

- Most small vendors do not maintain staff levels to handle "ad hoc" service requests for non-contract clients. If your processes are critical and a lengthy outage will prove costly, pay the monthly service fee or hire your own IT professional.
- Keep in mind, when a vendor first visits, they will be unfamiliar with your environment. If the scope of the project changes, you should expect the price to increase. Work with them to agree on a price that is reasonable to both of you.

- Notify your IT provider when acquiring new applications. Your provider can evaluate them to ensure they will integrate with your current operating system and existing applications. The time to discover an application won't work is not days before you are ready to begin installation.

- Notify your IT specialist when you receive notices from other service providers. For example, a notice about DNS changes from your Internet Service Provider is something they need to know.

Finally, as much as possible, use one provider to maintain your infrastructure. I once helped a client setup a wireless network, who later decided to buy a refurbished laptop from a different vendor. While setting up this laptop, the vendor changed the interface for the wireless network, leaving his assistant unable to access the network. The laptop dealer didn't consider how the office was set up. My client didn't anticipate that he would need to communicate these details to the dealer.

This illustrates what can happen when you involve multiple vendors. We understand your need to spend capital wisely. We also understand how too many cooks can spoil the broth. Your primary vendor likely has resources who already understand his processes and will work within his parameters to ensure their efforts don't compromise other work.

1. The website www.itminefield.com has links to respected strategic planners.

I.T. Doesn't Happen To ME... by Jaime Buckley

"I THINK WE SHOULD CHARGE AN EXTRA $10K FOR OUR PROPRIETARY I.T. SOLUTION SYSTEM!"

Chapter 3

How to Pick an IT Vendor

The task of looking for an IT vendor or a support provider rattles most small business owners. As a group, IT service providers tend to enjoy playing with technology more than talking to people. The "techno-speak" often sounds like a foreign language. When you call on them for troubleshooting, they generally want you to define the problem and then go away while they solve it. If they call on you, they present their services in terms of what the technology does rather than what your business needs. You don't leave these encounters with a warm and fuzzy feeling.

Another challenge is finding someone willing to provide IT services to a smaller firm (small in terms of number of computers, not revenues). Most IT service providers who start their practice supporting small businesses find they need to develop a more profitable secondary focus (e.g. application support and development). Consequently, they drop their basic support as the secondary focus grows.

As a group, it is a challenge to build a profitable IT support operation around small businesses. Why?

- Small enterprises tend to have a very stable environment. Once the infrastructure is established, only limited changes or updates are needed over time.

- Pen and paper are sufficient for many tasks, so process automation is not cost justified.

- Generally, small businesses prefer not to commit to service contracts. When something breaks, the small business wants help immediately. Until then, engaging in a support contract for expedited or preventive service doesn't seem worthwhile. It's a cost vs. risk debate just like paying for an insurance policy.

You want a relationship with a vendor who listens to you, understands your business and is looking out for your best interest. How do you find such a firm?

First, check with other businesses in your industry. Seek a provider who specializes in supporting your type of business. For example, a local provider might specialize in healthcare offices. Another one might support the small manufacturers in your area. These vendors will be most in tune with your business and will keep up with the technology innovations that impact you. Don't be concerned about the possibility that someone might give you a bad referral to hinder your enterprise. According to Degraffenreid we all make referrals to make ourselves look good. Consequently, even if you get a referral from a competitor, your competition will still give you their best referral to maintain their image.[1]

If you can't find someone who supports your industry in general, then get a referral from a business that has a similar number of computers and similar environmental structure. If you are an engineering firm you might talk to a graphics firm, but probably not a tax accounting office. A financial advisor could look at a small non-profit or an insurance provider.

In any case ask the service providers to come in for an interview. Ask for three references. You want to get the client's perspective on the level of support they are receiving.

Interviewing the Service Provider

Ideally, you want to be at least as comfortable with your service provider as you are with your personal physician. What qualities should you seek?

- Experience. How long have they been supporting businesses like yours? Will they provide you references?

- Stability. How many customers do they have? How rapidly are they growing their customer base? Will they be able to take you on and still provide quality service?

- Certifications. Look for certifications such as a Microsoft Certified Partner. These businesses have gone to the effort to meet Microsoft standards. Individual certifications for employees are important, such as a MSCE.

- Long-term relationship. Is this vendor a break/fix provider or long-term partner? The long-term partner wants to be proactive in preventing problems and presenting solutions for your business. They expect to spend time with you on a regular basis, whether you have a problem or not. The break/fix provider only expects to hear from you when things break.

- Good communication skills both listening and speaking. Do you feel like you have been fully heard, your requirements understood and your viewpoint considered? Do they present their ideas in terms you understand? Or do you feel most of what they say is lost in translation?

- Focus. Are they focused on your business needs or the technology and what it can do? Are they asking you questions about your business (such as the questions in Chapter 2 *Understanding Your Business*)? Or are they asking questions about processor speed or hard drive size?

- Connections. Very few small IT service firms can meet all the needs of a small business. With whom do they partner to provide full service support? Do they have a network of providers to fill in gaps in their support? Are they willing to talk to other vendors on your behalf?

- Transition plan. No one likes to consider the possibility, but what is going to happen to you if their business should fold? What is the process for ensuring a smooth transition to another provider?

At the end of the interview, ask yourself, "How do I feel about having this person/firm work with me?" If you're not comfortable, walk away. If you're not sure, wait until you have talked to their references to get a broader perspective.

Interviewing the References

Talking with references reveals a great deal about the service provider. The vendor wants you to get the best possible impression of his products and services. References may be brutally honest with you about the service they are receiving. A disconnect between the service provider's perception of his service and the reference's perception should raise a red flag.

During these interviews you will cover the same topics, but from a slightly different perspective.

- Experience. Does the provider find solutions quickly? Or do they spend a lot of time in trial and error approaches? Do you feel like your business is in good hands, or is your business their training ground?

- Stability. If the vendor has employees, do you get to work with the same few each time? Or is their staff like a box of chocolates, you never know what you're going to get? You need to get a feel for the vendor's ability to retain good talent. You do want some variety. If different employees work your account, it reduces the chance that all of their knowledge of your business will walk out the door. Each employee should document their work, so any qualified service provider can step in and handle your account.

- Relationship. Do the vendor's solutions come from a clear understanding of your business strategy? How much time has the vendor spent with you talking about business strategy? Is maintenance proactive? Does the vendor alert you to important technological developments?

- Communication. Have you had difficulties with your vendor? What was the problem? How did those conversations proceed? Were they easy, difficult or punishing? What was the outcome?

- Focus. Over time, does the vendor seem more interested in the technology than your business? How is this apparent?

- Connections. If the vendor outsources to fill gaps in their service, how satisfied are you with the other vendors? Does the vendor communicate with them on your behalf?

- Transition Plan. What systems are in place to make it easier to transition to a new vendor if it becomes necessary? Is all of the documentation about your environment up-to-date? Or are they withholding vital information?

- Disaster planning. If they are managing your servers or website, have they ever successfully tested their disaster recovery recovery plan?

- Anything else? Does anything stand out about their performance?

Two final questions to ask of the reference:

- Overall, how satisfied are you with your vendor's performance?

- What compels you to stay with this vendor?

Armed with these perspectives, you will have enough information to make an informed decision about whom to choose as your IT service provider.

1. Degraffenreid and Blandford, The New Art & Science of Referral Marketing

I.T. Doesn't Happen To Me. by Jaime Buckley
"ACCESS DENIED."
Copyright 2009 Jaime D. Buckley

Chapter 4

Document Everything

A new client called looking for support. Their first vendor literally "got religion" and moved away. Their current ad hoc support person was moving out of the building. They were changing broadband service providers and required some parameter changes to their firewall appliance. I walked into their office expecting the change to take 30 minutes or less. I asked the client for the firewall documentation. They didn't have it. Another group in the building had procured and installed the firewall. So, I went down the hall and asked. They didn't have the documentation either. I thought to myself, *OK, I don't really need the documentation. I just need the administrator ID and password.* I asked. They didn't know it. After calling their IT support, we tried several passwords they typically used but were not successful.

I suspect you're starting to get the picture. What should have been a very simple and inexpensive change turned into a time consuming operation: obtain documentation and firmware from the vendor; understand the process for resetting the firewall; return to the client; erase and reinstall the firmware; document the process, new settings and the administrator ID and password.

Anytime you allow someone to make a change to your network, servers, hardware or software, require them to leave you all documentation:

- What was done and why
- Original and new parameter settings
- Software and documentation CDs
- Changes in work processes
- Changes in passwords

Store all of your documentation in a fireproof safe or file cabinet.

Some service providers may consider you high maintenance, but hold your ground. If they don't agree up front to document their work, send them away. This provider isn't protecting your interests but their own, making it hard for you to switch to another provider.

In fact, the purpose of documentation is to safeguard your business. Employee turnover is inevitable. Even good service providers go out of business unexpectedly. A paper trail is the easiest way to help a service provider help you. Having information about your environment readily available simplifies problem diagnosis and resolution.

Asset Inventory

Whether you are an established business or a start up, you need an asset inventory. An asset inventory is a record of every piece of hardware and software your business owns. These inventories, along with the network documentation, are especially important if you need to rebuild any portion of the IT environment or collect on a claim to your insurance company.

You can purchase asset management software, create your own database, create a spreadsheet or rely on pen and paper. Any of these methods will work. Pick the one that suits you and stick with it.

A hardware asset inventory includes:

- Device manufacturer
- Device model
- Device serial number
- Device location
- Date in service
- Amount of memory
- Hard drive size
- Unit cost
- Administrator account name and password
- Software installed and version
- Maintenance service log which includes date, service performed, and any configuration changes.

A page in your notebook might look like the example on the following page.

Device Location:		*Receptionist desk*	
Manufacturer & Model		*Dell Vostro 200*	
Serial Number			
Service Tag			
Date in Service		*1/1/2007*	
Cost		*$1999*	
Account Name:		**Password:**	
Hard drive	*120G*	*Memory*	4G
Software Installed		**Version**	**Date**
MS Office Basic		2007	1/1/2007
Visio		2000	1/1/2007
AVG		7.0	1/1/2008
Quicken		2007	7/1/2008
Service Log:			

Date	**Service**
6/1/2008	*Upgrade memory to 4G*
7/1/2008	*Installed Quicken*
8/31/2008	*Email problem; Called Support*

Table: 4.1— Hardware Asset inventory sample.

A software asset inventory should include:

- Application name and version
- Application vendor
- Date installed
- License key or serial number
- User(s) assigned
- Cost

The Business Software Alliance provides links to free software tools to create the inventory in the Anti-Piracy section of their website: http://www.BSA.org.

Although this level of documentation might seem onerous or excessive, an up-to-date inventory provides a vendor with important information about your environment without them having to dig it up. It also saves you many headaches when dealing with your insurance adjustor.

If you are keeping your inventory in a notebook, periodically scan a copy into a folder and keep a back up for disaster recovery purposes. If the inventory is stored online, make sure it is in a folder that is backed up regularly.

For each device, you should have a file folder, small box or Ziploc bag to hold all software and other documentation. Keeping all of the information for a device in one container will simplify problem diagnosis and resolution.

You are allowed to make a copy of the software discs for backup purposes. Store the originals in a safe place, preferably away from your facility. Be safe, not sorry.

If your computing environment has been operating for a while without an inventory, take the time to create the asset documentation or have it created for you. Don't wait to discover you lack the proper documentation when you need it and can't reach the business that set up your systems.

Adding a New Employee

When you add an employee, several housekeeping tasks must be performed: fill out paper work, prepare an office for them, assign them a computer, and in many cases, grant access to shared applications.

On a separate piece of paper or a new page in your IT Manual, write down all the tools a new employee needs to perform his or her function. The list may differ for each position in your business. Below is a head start for your list.

- Computer
- Email account
- Printer or access to a shared printer
- Access to a shared data server
- Access to specific shared applications
- Access to internet applications
- Cell phone, smart phone or PDA
- Special software
- Training on software or company processes

Each item on your list potentially represents a process that should be documented, particularly if it is outsourced. Don't let turnover or changing vendors stop your business in its tracks!

Update the software and hardware asset inventories each time you add a new employee. Make sure you have purchased sufficient software licenses to account for the new employee. Compliance with software copyright law is your responsibility.

When you are well-organized, prepared for your new employee, and able to facilitate their smooth transition into the company, you get an additional benefit: Image. Your new employee will be impressed, and word will get out your business is well-run.

We Have a Problem

While the computing environment in most small businesses is fairly stable, changes occur. An internet service provider changes DNS servers. Automatic maintenance updates occur to your computer overnight. A virus slips past your security and infects your system. Any of these can change the environment to the point that critical processes quit working.

Once again, documentation is the key to properly handling the problem. First, document your view of the symptoms. Then, have your service provider document the actual diagnosis, the solution implemented, and any changes to your work process that will result. Keep a copy of all documentation for your records. Maintain a problem log, noting when the problem occurred and how it was resolved referencing the documentation. It only takes a few minutes and it will save you hours of backtracking.

Your service provider will track this information internally, yet you need to keep the same information on site in case you change service providers.

GRANTLAND®
I NEVER THOUGHT OUR BUILDING WOULD CATCH FIRE LIKE THAT.
Copyright Grantland Enterprises. All rights reserved.
IS OUR DATA OKAY?
OH YES, BOSS. EVERYTHING'S ON DVDs AND TAPES.
IN A SECURE PLACE?
ABSOLUTELY. THEY'RE ALL IN MY OFFICE...
www.grantland.net
CRASH!
SAFE.

Chapter 5

When Disaster Strikes

While attending a conference, a young IT savvy executive left her laptop in her hotel room. She knew the hotel cleaning crew by name and didn't expect a problem or feel her assets were in danger. Unfortunately, a man posing as an associate entered her room while it was being cleaned and stole her laptop. She shed no tears that night.

After all, she thought, I have a backup.

After acquiring a new laptop, she proceeded to restore the necessary files. Then she discovered that parts of the backup didn't work and many files were missing. The weeping began.

After 45 days, the ordeal was finally over: 30 days to settle the claim and acquire a new computer and 15 days to recover the available data. Some files were lost forever.

You Too Are Vulnerable

When small business owners think of a disaster, they tend to think of natural disasters that don't strike often, yet are impressive and extensive in their impact: tornadoes, hurricanes, earthquakes and floods. In IT, a disaster is ANY event that prevents a business from executing critical processes that keep the business viable. Disasters expose the vulnerability and weakness in our IT environments.

Many disasters are neither highly publicized nor visible. They might be triggered by trivial events, yet they are more common and can be more devastating than a natural disaster.

One television commercial shows a vacuum cleaner falling over and setting off a chain reaction that eventually activates the sprinkler system destroying all the electronics in the office.

A hacker gains access to the network and takes it down along with the servers or they may unleash a virus that infects, deletes or steals critical files. A construction crew severs a fiber optic cable disrupting communications for 4 days. A new hard drive starts to whine in the night and by morning the drive is destroyed. A critical supplier to your business suffers a disaster and is unable to recover. A production company loads their computers and monitors onto a pallet for transport, and under the cover of night, someone walks off with the pallet.

Some disasters are as simple as spilling a glass of chocolate milk or a sticky soda on the equipment or accidentally deleting an important file. When it comes to IT, Murphy's Law is definitely at work: If anything can go wrong, it will.

Then we have a previously unimaginable event combined with incredibly poor judgment. Many of my consultant friends had clients based in the World Trade Center complex on September 11, 2001. One told of a company that had terminated a data backup contract and brought the backup tapes to their WTC offices where their servers were located. The servers and the backups were all destroyed in one day. A new contract had not been signed and new backups had not been initiated.

Disaster Recovery Planning

No one is immune. For most small and mid-sized businesses it isn't a question of if a disaster will occur, but when and how bad will it be? How long can your business afford to be down before you are forced to close the doors?

"Nearly half of small companies that lose business-critical data in a disaster do not reopen after the loss...90% are out of business within two years."[1]

From this statistic, it is clear that most businesses are not prepared. Recovering from a disaster takes more time than most would predict and often more time than they can afford.

Disaster recovery or business continuity planning is about risk mitigation; to examine the potential threats and seek ways to either eliminate the risks or reduce delays. Disaster recovery planning begins with three questions:

1. Which processes are critical to your business?
2. What are the technology components upon which these processes depend?
3. How long can your IT environment be unavailable before the business starts to suffer?

Armed with this awareness, you then imagine the events or components that could cause a critical process to fail. Critical components include:

- Files (due to virus or accidental deletion)
- Hard drive
- Server or other hardware
- Hosted application
- Network/internet or telecommunications
- Special purpose hardware
- Electrical power
- Physical facility
- Parts and product suppliers

Recovery time will be impacted by several possible factors. Consider the following areas:

- Plan details: Do you have a detailed plan that specifies exactly how to recover a process or component? Could someone unfamiliar with your environment execute the plan? Have you outlined alternate methods for executing the process that are not technology dependent?

- ♦ Hosted applications: If your website or other critical application is hosted by a service provider, have you verified that they have a disaster recovery plan for your application and that it has been successfully tested? The results of an SAS 70 audit usually will indicate their level of preparedness.

- Insurance: If your facility is destroyed or several computers are stolen, do you have sufficient insurance coverage or funds in the bank to cover replacement costs and leasing a new facility?
  - ♦ The insurance company is going to want a list of assets lost along with the original purchase date and price. It can take six weeks or more to settle with the insurance company. Keep your asset inventories up-to-date to expedite the process.

- Power availability: Hurricane Katrina (2005) and the Oklahoma ice storms of 2007 and 2008 left many without power for several weeks. The speed of power restoration is often determined by the number of physical bodies the power companies have available to assign to the task.

- Network availability: If your facility is destroyed, how long will it take a telecomm provider to establish new phone and broadband service? A 14 to 30 day wait for a new install is not uncommon.
  - Does your provider have a redundant network to limit your downtime in case of a cable cut? Do you need two different lines from two different providers?
  - Looking internally, how long did it take to create the network in your building? Expect it to take just as long the second time.
  - If you have experienced a regional disaster, how long will it take your IT service provider to get to you?

- Hardware availability: Identify your critical hardware components. Do you have special hardware that is not available at the local computer store? Can you afford a 14 to 30 day wait to receive a replacement? If not, keep a spare on hand.
  - If it has been a few years since you last purchased new computing technology, you will likely be ordering the newest by default. It will take you or your service provider a week or two to design a system or environment that is functionally comparable to what was lost.

- Software availability: How up-to-date are your critical software components? Did you store the original installation disks somewhere safe away from your facility? If not, you will be forced to install a new version which will likely cause more problems than you realize.
    - Recognize that your old software may not work with a newer operating system. Even if you have the original discs, you may still have to upgrade your program.
    - If it has been 4 or 5 years since your last upgrade, you may have to go through 4 or 5 upgrades to get the database to work with the new software.
    - If the software was customized (e.g. special screens, forms, queries or programs), that customization will need to be repeated. Do you have a vendor available to take on that task? How long did it take the first time? Allot the same amount of time for the replacement.
    - Remember, you have a right to make a backup copy of all installation discs for recovery purposes. Do it, and store the originals offsite. This does not constitute software piracy or copyright infringement.

- Backup availability: How long will it take to retrieve your backup files online or have them delivered? Are they overseas or somewhere in the U.S.?

- Backup frequency: If a critical file changes daily and it is backed up every night, then restoring the file will bring the data up-to-date except for the transactions lost on the day of disaster. A less frequent backup will result in data loss.

- Vendor availability: If you don't have in-house IT support or a contract with an IT support vendor, who will help you recover? Most small business IT service providers don't maintain the bandwidth to take on a recovery project (small or large) on short notice and won't even try to fit you in without a prior arrangement. How long will it take you to find help?

You can't eliminate all of the delays. However, many of these factors are within your sphere of control or influence. For example, security policy can help protect against viruses that delete files or take systems down. An adequate data backup policy will reduce delays caused by an accidental file deletion.

The goal is to reduce the length of the longest delays to something your business can survive. You may look at some of these factors and decide that you cannot afford to plan for them or recover from a particular scenario. You may decide in the event of a disaster, the plan is to simply shut the business down. That's a business decision only you can make.

In a perfect world, disaster recovery would be painless, invisible and automatic. Your business would not skip a beat. Unfortunately, that solution currently does not exist and implementing even a 90% solution (a backup data center) is cost prohibitive for most small businesses. As you can see, being down for 12 weeks is well within the realm of possibility. That being the case, the key to surviving a disaster is planning and preparation.

"The Plan" should be kept in hard and soft copy in several safe places, away from your site. Place a copy in a folder that is part of your data backup. Give a copy to the IT service provider as well as employees that will be involved with recovery. As with all plans, they should be reviewed periodically, at least annually. Update the plan each time your IT environment changes (new process, vendor, or application). Finally, test your plan annually. You don't want a disaster to occur and then discover that your plan failed or that your backup files were inaccessible, incomplete, or unreadable.

Data Backup

How much data will you lose in a disaster? It depends on the type of disaster and the quality and frequency of your backups. Disciplined backup practices can limit the amount of critical data loss.

Focus first on the mission-critical processes. The files and applications associated with these processes are mission-critical as well. When thinking about the files, think of them in three broad categories: mission-critical, important and archival.

Mission-critical files are those that are absolutely necessary to the process or application. The business cannot function without them. Mission-critical files might also include those that would result in fines if they should become unavailable. For example, are you required to keep emails or royalty payment records? If so, what happens if these files are unrecoverable?

Important files and applications are those that are not required immediately, but you will need to restore them soon. These might include marketing materials, HR applications, CAD/CAM drawings or employee work files.

Archival files are those that are kept long-term for compliance purposes. These would include closed client files or files kept for regulatory purposes. These files do not change much once they are created.

Keep as many backup versions of a file as you need to minimize the recovery effort but not so many that the cost for backup skyrockets.

In your IT Manual, list your processes and their associated applications and files. Document file locations and categorize them based on their importance to your business. Assign a value of 1 to a mission-critical process and 2 to an important process. Assign a value of 1 to a mission-critical file, 2 to an important file and 3 to an archival file.

Put some time and thought into organizing your files. Create policies that define file naming conventions and storage locations for use by employees. Storing files in a few central locations will simplify the backup and recovery processes.

Finally, establish a data backup policy and communicate it to the employees. This policy includes:

- Identification and definition of mission critical processes, applications and files.
- Identification and definition of important files, processes, and applications.
- Identification and definition of archival files.
- Frequency of backup for each category of files.

- Organization and placement of files on servers, desktops and laptops to facilitate backup.
- Organization of email data stores to prevent loss of communiqués.
- Definition of protocols for adding processes, applications and files to the recovery plan.

Make it easy for employees to comply with the policy, and your critical data will be well protected.

Data Backup Options

Most small businesses choose to purchase hardware appliances for data backup at the local computer or office supply store. Many of these systems will initiate a backup on your schedule. The disadvantage is that often the appliance or the output tape or CD is kept on site. If the building and computers are destroyed so is the backup. You're screwed.

Another option is to contract with a service to backup your data over the internet and store it at a remote location. The service provider can initiate the backup on a preset schedule or you can control the schedule. Some of them will perform an incremental backup; only new files and files that have changed since the last backup will be backed up. These offsite services are advantageous, since offsite storage protects your data if the office burns down and your computers melt. Your data is safe elsewhere.

Choosing a Vendor

Many new small firms now provide data backup. Some are "mom and pop" operations that have simply added a "new" service to their IT business, possibly as a re-seller of another provider. Many offer data backup exclusively. Very few offer full-service disaster recovery solutions.

When choosing a vendor to backup your data over the internet, consider the following:

- How secure is the backup site itself? Is it in the garage of the owner's home? Or in a physically secured facility?
  - ♦ Serious vendors will have at least two physical facilities or data centers separated by a great distance so as not to be overtaken by the same disaster. Data is replicated in both facilities which function as "hot sites" for each other, taking over operations when needed.
  - ♦ Entrance to these facilities will be tightly controlled to limit access to data.

- What is the distance between the backup site and your location? If it is only a mile down the road or in the same town, your data could be lost in a natural disaster (think hurricane Katrina).
- Are they encrypting your data before it leaves your facility?
- Do they allow you to determine the number of versions of a file to keep?
- If your business requires it, do they have the ability to recover email messages?
- Do they have the ability to backup a file or database while it is open? Or do you need to adjust your processes to shut programs down for the backup to run successfully.
- How do they handle deleted files? Do they maintain a backup until told otherwise? If not, how long?
- Do they offer a variety of storage options? If the amount of data is under 100GB, variety may be less important. However, for large amounts of data, it will save you money to evaluate your files and their utilization in order to backup and store them properly. Some vendors will automatically move a file from one type of storage

to another based on its usage unless you specify otherwise. Generally, the vendor should support three types of backup:

- ♦ Online: Online backup provides the fastest access to your files regardless of how recently they may have been updated. Online backup should be used for files that are critical to your operation.
- ♦ Near-online: A near-online or nearline backup will generally take slightly longer to restore. However, it is less expensive than online since the requirements for access are not as critical.
- ♦ Offline: Offline backup is for those files that are primarily kept for archival or regulatory purposes. Generally, only one copy of these files is kept in a secure storage facility.

- Does the vendor notify you of backup failures?
- Does the vendor manage backups that you create? For example, you may initially create a backup hard drive for each computer and server in your facility. Will they store and track these devices for you?
- Does the vendor support a full server backup? Recent advances in server virtualization are leading to flexible and faster recovery solutions.

When combined with an appropriate backup solution, this represents a powerful alternative to backup only or a very expensive "hot site".

- If you should need to change vendors, for any reason including their business closing down, what is the process for retrieving your backups, transferring them to a new vendor and ensuring the old ones are destroyed and not available to the old vendor?
  - ♦ A reputable vendor will, for a fee, copy your data to appropriate media and ship it to your new vendor and then destroy their copies.
    - ∴ While changing vendors, do not bring your backups into your facility with your live data.
    - ∴ Do not terminate your backup contract with a vendor until you have a new contract negotiated.
  - ♦ If the vendor won't discuss this process with you, then walk away. You own your data. You have the right to change vendors whenever conditions warrant.

Disaster Discipline

In the earlier examples, both the executive and the WTC business took disaster preparation seriously. They had a plan and backup copies of their data. However, like many, they were lulled into a false sense of security and never really expected anything bad to happen. When considering the effort you will expend on disaster preparedness ask yourself, "How long can my business afford to be down before I am out of business?"

If you can't afford to be down for an extended period of time, then be prepared to invest more money on disaster recovery planning. It will be money well spent. Think of it as an insurance policy. You may never need it. But if you do, you're covered.

The important lesson to learn from the examples: BE DISCIPLINED!

- Keep critical files organized in a few folders on the fewest hard drives possible.
- Keep the backup physically separated from the real data, offsite in a safe place.
- Perform your backups on a regular basis.
- Review the error messages created during a backup. Make sure no critical files have error

messages. If they do, correct the error.

- Test your backup periodically by restoring a few critical files. Better to discover a problem while you still have access to the original files. Don't wait until you need to restore data for real. According to Gartner Inc., "34% of companies fail to test their backups. Of those that do, 77% have found tape backup errors."[2]

- Test your recovery plan periodically, correct errors and retest until you perfectly execute the plan. An untested plan will likely compound a disaster, not save you from it.

1. CPA Technology Advisor, *The Most Common Uncommon Disasters*, November 2006.
2. Cited at http://www.corevault.com

I.T. Doesn't Happen To ME... by Jaime Buckley

"Lose your password for your passwords again?"

Chapter 6

Passwords & Security

For years, the password was an easy and convenient way to ensure security and protect information. Unfortunately, today the four-digit password protection is no longer adequate. Any desktop hacker can crack a four character password.

Today, password creation is quickly becoming an exasperating experience. Every application requires something different. Some will allow a six character password. Some will require eight or more characters. Some will allow all letters or all numbers. Others will require you to have at least one letter or number. Some will allow real words. Others won't.

How many different passwords do you have? It is almost impossible to start out with a basic password you can alter as changes are required. No wonder passwords give us a headache!

Password Guidelines

Passwords are supposed to be kept secret. Writing them down is discouraged. Yet what choice do you have? You've already seen one example of what can happen when a password is lost and you can't recover it.

You need to maintain some type of password documentation just in case the person maintaining your environment is no longer available. This is especially true for non-profits. People become involved and are around for years. Then life happens, and they move away. Here are some guidelines you can follow:

- For personal passwords to business accounts and web based applications, such as banks, Dell, E*Trade and Amazon, don't worry. These applications usually have an easy way to retrieve or reset passwords. Grant power of attorney to someone, so they can access important accounts in case you die or become incapacitated.

- If a web-based application will allow you to make a purchase without creating an account, do so. One less password to remember.

- For passwords known only to you and your IT support team, choose an 8-digit base that you can easily recall and add one more character to it. Write down all the applications that use the password (not the password). When you are forced to change a password, change all of them by altering the last variable character.

  - ♦ If you keep a list, on one piece of paper write the application and the additional character. If you have to write down the base, do so, but make sure you store each piece of paper in a separate secured place.
  - ♦ Another alternative is to keep the passwords in an encrypted file that is password protected and backed up. Just remember, if you forget that password, you have a problem.

- Never reveal administrator passwords to your employees. Computers and most applications come with administrator ids. They allow you to create user accounts for your employees that will allow them to log-in to an application without administrator privileges. This is to protect you in case an employee leaves or is dismissed.

- If you have files that need more security than others, place them in a unique shared folder and grant access only to essential users.

Security Packages

Everyone needs basic protection against viruses, spyware and malware. I don't have a strong opinion on which one to use. Some folks swear by Norton and others by McAfee. Still others curse both of them.

Microsoft Windows® provides a software firewall. You do not need to purchase a separate one. If your security package also contains a software firewall, then you generally need to turn one of them off (usually the Microsoft firewall). Running both could cause you problems.

If you purchase a hardware firewall appliance, you don't necessarily need the software firewalls though they can work together when configured correctly. A hardware firewall is adept at preventing inbound attacks. A software firewall proficiently stops outgoing problems, such as email worms.

Security Policy

You can find many white papers and opinions on security policy using your favorite search engine. Some security policies are mandated by the government (e.g. HIPAA, SOX). You are responsible for knowing security requirements with respect to your industry. Here, we are more concerned with the policy that will protect your computing assets.[1]

Smaller business owners tend to be more lax in their security practices, thinking: *My employees wouldn't do that!* Yet most of the computer problems in the office computers are caused by the employees.

How? By allowing employees:

- Unrestricted access to the internet (e.g. gaming, shopping, movie and porn sites)
- To treat company computers as personal computers
  - Processing of personal emails
  - Download and installation of non-business software
  - Personal shopping
  - An office babysitter for their children
- To circumvent security policy
  - Attaching a personal computer to the network
  - Creating a personal wireless access point in the office
  - Storing company work product in folders that are not backed up
  - Taking company files home on removable media (e.g. thumb drives, CD's)

Security policy should not be viewed as a ball and chain. A slip in any one of these areas can cost you productivity and revenues. Many websites are not harmful, but it only takes one to unleash a nasty virus or Trojan horse (spyware) on every computer in your office. Losing only a few critical files can sink your whole enterprise.

Acceptable Use

An Acceptable Use Policy defines which activities are allowed and which are forbidden as well as the consequences of noncompliance. Consider carefully what you want your employees to be doing with company computers on company time and what you don't want them doing. Do you want them playing internet games, viewing movies, shopping for lingerie or a new job? Do you want their children tapping away at the keyboard?

Set your parameters then talk with your service professional about the security policy you would like to create. Communicate your expectations so he can take appropriate measures to lock down your network and your computers and prevent unauthorized use and access.

You can employ hardware appliances and software to block access to websites and reduce spam. Administrator settings can prevent unauthorized software installation or the use of USB flash drives (which can accidentally introduce viruses).

Next, document your policy, and communicate it clearly to the employees, both informally and formally. The formal communication may be delivered in a companywide meeting and at new employee orientation. Present each employee with a copy of the policy and have them read it and sign it, declaring that they understand the policy and agree to comply.

Also, initiate informal "water cooler" conversations. Indicate you are aware of their efforts to comply with the policy. Ask if the policy is hampering their work or creating expensive inefficiencies.

Explain that their compliance protects the business, and non-compliance will hurt the business financially or even inflict damage that would result in closing the business. If the policy seems reasonable and can be followed without hindering work or forcing employees to be overly conscious, they will comply.

Most importantly, once you decide on a policy, be prepared to enforce it equally across the company. The rules that apply to the employees also apply to

visitors, upper management, including you and any relatives you may employ in the workplace. Playing favorites will quickly erode your ability to enforce any policy.

Changes in your business may necessitate changes in your policy. Be sure to document and communicate the new policy.

If you wish to provide more generous access for personal business, consider placing one computer for personal use in a central location. It should be blocked from sending inter-office e-mail. Be sure it has zero access to any other computers or servers in your network. Make sure it cannot burn CDs, DVDs or write to a USB flash drive, lest it carries a virus back to the main computers. An isolated computer limits your risk.

Remember, once one computer is infected, it doesn't take long for all of them to catch the bug. Without a security policy, be prepared to spend more money on eradicating viruses, malware and spyware or potentially enacting your disaster recovery plan.

1. Clearing the Minefield will cover other specific types of security policy which your business might need.

HOW ARE YOU DOING WITH THE NETWORKING PROJECT?
SO FAR, SO GOOD --
I'VE GOT THE FIRST TWO COMPUTERS TIED TOGETHER,
AND I'VE JUST STARTED ON THE THIRD ONE.
Copyright Grantland Enterprises. All rights reserved.
www.grantland.net
1612

Chapter 7

Servers & Networks

The task of deciding which computers and software to purchase is often daunting. Add a conversation about servers and networks, and it is enough to put the small business owner into a catatonic state.

Many books have been written to cover these topics in great depth. If you're interested in the details or "how to", please visit your favorite bookstore. My objective is to help you communicate your needs to an IT service provider based on your business strategy.

Do I Really Need a Server?

In the old days, having a server meant buying the biggest computer with a special operating system and the most capacity you could afford. A network doesn't necessarily require a special computer with a special operating system. Today, you can actually create a server from a decommissioned desktop or laptop by simply adding more memory and possibly hard drive capacity. If you are planning to share anything between two or more computers, you need a dedicated server. The real question is: can it be a peer (an ordinary computer) in a Windows® network or does it need to be a true network server with a network operating system?

To determine the best fit, you and your service provider will consider the following questions:

- What do you need to share? Files, printers, scanners? Applications?
  - Printers, scanners and other hardware don't generally require much special consideration. A server that acts as a peer in the network will be adequate.
  - Let your provider know which applications you are planning to share so he can check system requirements. The main concern is the integrity of a database that is accessed by multiple users. Does the application require a network server or can it function in a Windows® peer-to-peer network?

- How many computers are likely to have access to the network at the same time?
  - If 10 or fewer, then the sharing capability in the Windows® operating systems will likely be adequate.
- When do you project more than ten employees on your network?
  - Don't be optimistic, be honest. If it won't be for several years, you probably won't need to invest in a special network server.
  - If you require a network server you will pay a license fee for each user on your network. Your costs will be higher.
- Do you plan to host your web site? Email? How heavy do you expect traffic to be? What level of availability do you expect or need (e.g. 24/7)?
- Do you have any special security requirements? This potential requirement will impact the server and network configuration.
- Are any of the shared applications resource intensive, using lots of memory, or bandwidth such as engineering or graphic design?
- Are any of the files being stored on the shared resource likely to be large (e.g. engineering drawings, high resolution graphics)?

If you answered yes to the last four questions or have more than 10 computers sharing in the network, then you probably need to consider using a network server with an operating system optimized for networking, security and high availability. Otherwise, the basic file and printer sharing provided by a Windows® peer-to-peer network will be adequate.

About Networks

Network terminology can be intimidating. You have hubs, routers, repeaters, gigabit switches, modems, firewalls, filters, firmware, wired and wireless networks, twisted pair and coaxial cables. Are you ready to pull your hair out? That little list doesn't even begin to cover all of various appliances you can add to your network.

Still, if you can grasp the basic network concepts and understand the needs of your business, you can help your service provider make the best decision possible for your enterprise. The network is an area where you can spend excessively for little or no additional value. You can also skimp on hardware to your harm.

To demystify networks a bit, consider that a network serves three purposes.

1. Facilitate communication between computers within an office or a company.

2. Facilitate the sharing of common devices (e.g. a single high quality laser printer versus 10 little inkjet printers) and common files.

3. Facilitate access to the outside world.

You actually have two networks, though we often think of them as a single unit. The first network is the one beyond your office. You have little or no control over it beyond choosing a broadband service provider. Your service provider's main concern is the amount of traffic coming into and going out of the office and the availability requirements.

Does your office use the network primarily for e-mail and surfing the web? If so, then you probably don't need to pay for a dedicated T1 line, which provides fast, but expensive access. On the other hand, if your business has multiple locations or you are hosting your own website, and availability and access speed are important, you might want to procure more bandwidth or even have a secondary line.

The second network is the one within your office. This network can be wired or wireless. Without internet access, your wired network would look something similar to figure 7.1.

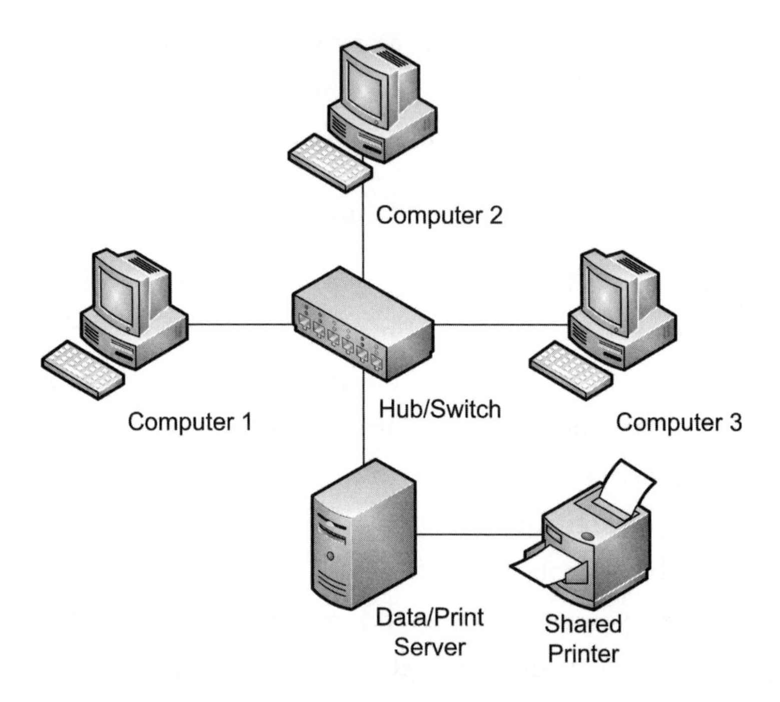

Figure: 7.1—Simple wired network

Each computer requires a network interface card (NIC) which connects to a cable (invest in good cables for your network). Cables run from each computer to a hub or switch which is likely housed in the telecom or central wiring closet. This closet contains all of your wiring, both computing and telephony. It should be secured from general access. If you can't have a wiring closet, place the switch or hub in a central room with your "server" and other shared devices.

When you add the internet to your wired network, the picture looks something similar figure 7.2.

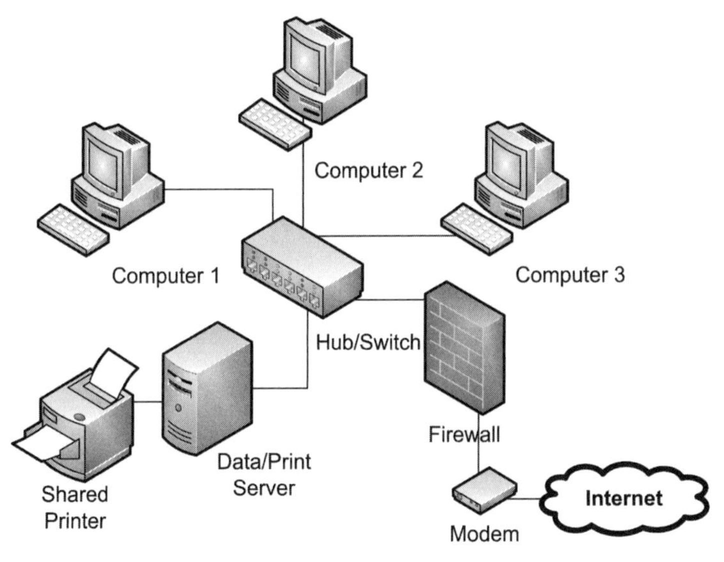

Figure: 7.2— Wired network with internet access

A firewall appliance has been added to protect the network from those who would try to hack in and take it over. It must sit between your network and the outside world. Without a firewall, a hacker can discover your network and compromise it in a matter of hours. One advantage of the wired network is that usually an intruder must gain physical access to your facility to connect to your network. If access to a facility is controlled, it is likely safe from intruders.

Another option is to have a wireless network within your office. The main advantage of a wireless network is convenience; no need to drill holes in the wall to run cables from each computer to the switch. Each computer must be equipped with a wireless network adapter (similar to the NIC in a wired network) which will either be built into the computer when you purchase it or added separately later.

Instead of a hub or switch, you will have a wireless access point (WAP) or access point(as seen in figure 7.3). These terms are synonymous and may be used by your service provider interchangeably. If the computers are more than 300 feet apart, you will need more than one.

An access point serves as a central connection point for the computers with a wireless network adapter. If you already have a wired network, the access point can be connected to your switch to allow the wireless and wired networks to co-exist (as seen in figure 7.4).

Some access points include features that allow them to function as a switch for a wired network, as well as a firewall, to permit you to connect to the

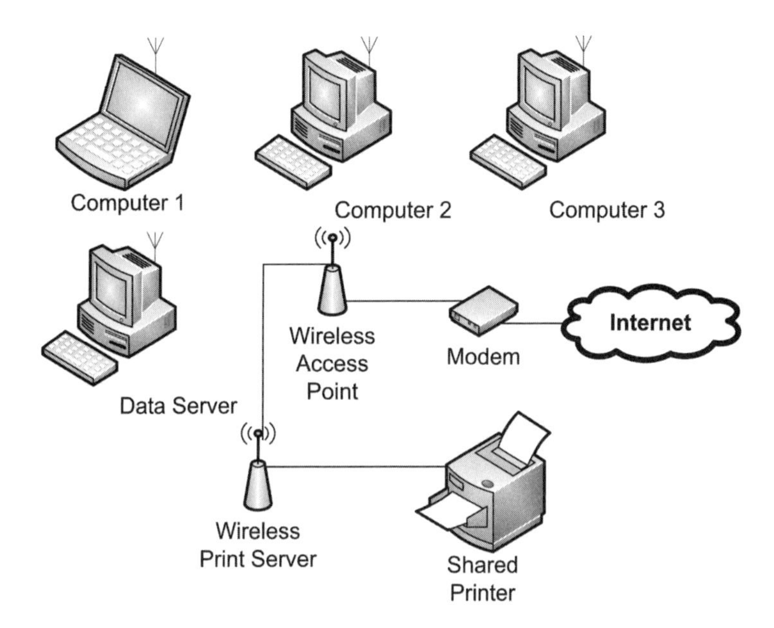

Figure: 7.3— Wireless network

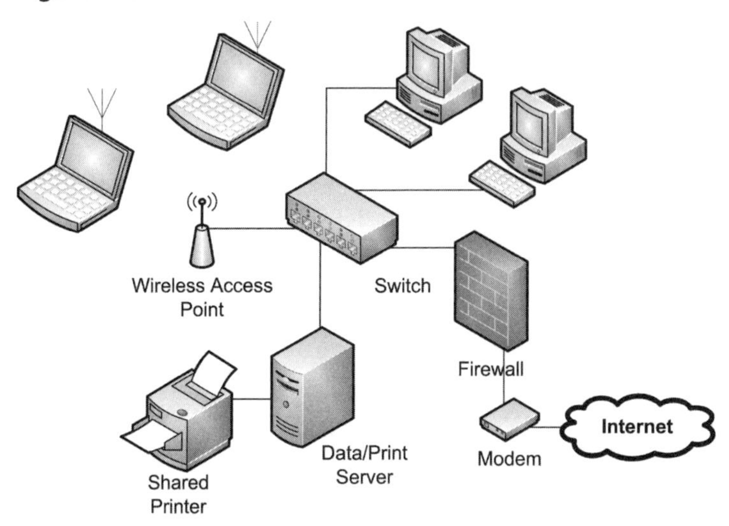

Figure: 7.4— Wired and wireless network

internet. Your service provider will determine which type of access point will meet your needs.

The main disadvantage to using a wireless network is security. While it is possible to have a secure wireless network, there are more opportunities to breach a wireless network if it is not installed correctly.

Since a wireless network communicates through space via radio waves, an intruder needs only to be within 300 feet of an access point to receive a wireless signal. Some intruders start out relatively harmless (cheapskates who don't want to pay for broadband and eavesdroppers who just want to prove they can) but their motives can quickly turn malicious.

To ensure your protection, the service provider needs to change the default values in four areas:

- Administrator ID and password. Store them in a safe place and don't lose them! Most access points have a web based administration interface that is easily accessible over the internet. If the default is in place, they can gain access and lock you out.

- SSID. The SSID identifies the network to client computers. All of the major manufacturers have a default that is well known and published on the internet.

- SSID broadcast. Broadcasting the SSID makes it easy for you and friendly computers to find the network; however, it makes it very easy for the intruder. Disable the broadcast.

- Encryption keys. Each access point uses some encryption method (usually WEP or WPA) to secure transmissions. The easiest way to change the key is to come up with a passphrase (similar to a password) and allow the software to generate new encryption keys. The passphrase is the information you will share with friendly computers to allow them to access your network. Their computers have the same encryption methods and will generate the same encryption keys you did.

If security is a serious issue, you should probably choose a wired network over a wireless one.

Last Word

I can't emphasize it enough, make sure you are the keeper of the documentation for your network. Once your network is up and running, the service provider(s) should turn over the following:

- Documentation and discs for each hardware appliance installed on the network.

- Documentation for administrator ids, passwords and processes for accessing the administration software. Do not lose the id or password. As an earlier case showed, it can be costly!

- Documentation for parameters (including passphrases and IP addresses from your broadband service provider) that were changed as part of your installation.

- A network map showing the appliances and their relationship to the computers and other shared hardware.

- If a switch or hub is used in your network, then you need documentation about which office is plugged into which port on the switch. The documentation can be tagged on the cables or written on a piece of paper.

If you don't have the documentation, get it! It is your responsibility to protect your business!

GRANTLAND®

OKAY, YOU'VE GOT $30,000 TO DEVELOP A SOFTWARE PROGRAM TO MONITOR OUR PRODUCTION.
GREAT!

ONLY $1,500 TO GET DAILY REPORTS? DO IT!
GO AHEAD WITH $2,000 FOR A QUALITY CONTROL FEATURE.

WORK STATION ANALYSES FOR $5,000? WE'LL TAKE IT.

HERE'S THE SOFTWARE PROGRAM. IT RAN A TAD OVER BUDGET.
$225,000

Chapter 8

Your Website

The effort to build a company website is not to be taken lightly. Some business owners dive in headfirst and spend several thousand dollars to get a site up quickly only to spend several thousand more to redo it. Often the conversation goes like this:

Client: "We took out a second loan to build the site. Now we're out of money and the site isn't ready. The developer isn't willing to continue the work on a contingency basis. We need to raise capital to finish it."

Consultant: "Oh. That's interesting. How much have you spent? How was the payment structured?"

Client: "Well, they charge by the hour and we've spent over $40,000"

Consultant: "OK. Did you and the designer agree to what would be delivered by a specific date?"

Client: "Yes."

Consultant: "OK. Then what happened?"

Client: "Well, we thought we should incorporate Web2.0. One partner wanted this feature. Another partner wanted that feature..."

After they go on for a bit, the consultant asks, "Do you realize what happened?"

That question is answered with a blank stare. They know what they have said, but they don't realize the impact. The conversation continues.

Consultant: "Who controlled the funds?"

Client: "We did."

Consultant: "Did the others speak with you before requesting a change from the designer?"

Then the light bulb turns on and they see the error in their ways. They failed to curtail changes to the original specifications. They neglected to create a process for partners and vendors to communicate desired changes and get approval within the budget. Consequently, they spent a lot of money on "nice to have" features and forgot about the central goal: to create a functional website.

The overall price tag isn't the problem. Many people have ambitious goals and high expectations for their sites, and this costs money. But failure to manage the project is a concern that needs to be addressed.

Generally, these situations arise when business owners engage enthusiastic, but probably inexperienced, website designers who extol the advantages of a dynamic website, with the "If you build it, they will come" pitch. The owner is dazzled by dreams of fantastic results, and they walk off arm-in-arm into the sunset, only to end up suing each other for breach of contract in the end.

Before You Begin

Fortunately, you can avoid this nightmare. Before you spend a single penny or sign a contract, take the time to understand your target markets and their use of the internet, your marketing plan and the role the site plays in your business strategy.

Consider the following questions:

- What is the purpose of your site? How does it fit into your overall business strategy? Possible answers include:
  - Create community: provide a place for people with like problems and interests to share information and solutions.
  - Establish credibility by providing a site with basic information (a glorified yellow pages ad).

  - Marketing: make your target audience aware of your products and services.
  - Sales:engage your target market in the purchase of your products and services.

- What results do you expect to achieve? This is your measure of success and it provides the designer with additional information about your ambitions. Your desired results should be stated as a SMART goal:
  - Specific
  - Measurable
  - Attainable
  - Realistic
  - Time bound

- How do you intend to drive traffic to your web site? What is so unique about your site or business that people will visit your site out of the millions of sites on the internet? Building a site does not guarantee traffic.

- Once you get them to your site, what is your target market likely to do? What do you want the visitor to do?

- Why will they return to your site? If they only visit once, you're not making the most of the opportunity.

- Finally, how much are you willing to pay to achieve your desired results? You don't necessarily need to reveal this information to your prospective vendors. Establishing the value you place on the results helps define the project budget. Keep in mind, a low budget will limit your options and, often, your results. Experienced developers and designers with verifiable client results can command $10,000 - $25,000 or more however, their solution will fit your business strategy.

The more you know about your target markets, the better you can direct the web designer's efforts. If you hire a designer before you are clear, you leave a lot open to the designer's imagination. Your project could turn into a "money pit"; a black hole sucking up all of your capital.

Picking a Designer

In the internet's early days, "real" web designers knew HTML, JAVA etc. but did not understand marketing or the psychology or mechanics of driving visitors to a site. With the proliferation of code generators, templates and do-it-yourself guides, just about anyone can create a web site and claim to be a web designer.

Whether a stay-at-home mom or the kid down the street or an off-shore designer in India, web designers all make similar claims concerning their work and results. Some designers will want to showcase their artistic ability or their coding prowess with "special effects", but very few of them have a strong grasp of marketing. Caveat Emptor! (Let the Buyer Beware). If all you want is a basic web presence to establish credibility, any of these people will work fine for you. Some of these "designers" will use templates, open source software or shrink-wrap solutions to provide online stores and shopping carts. It may cost less, but you might be vulnerable to security issues and your results may suffer.

If your ambitions are greater (as expressed by the results you want to achieve) you need a team that understands marketing and has a proven track record of employing solutions and strategies that will improve your placement in the search engine results.

A professional team will use search engine friendly solutions rather than generic shopping carts or affiliate programs. A full-service design firm will include members who understand branding, artistic design, user interface design, or they will have easy access to those talents.

Over time, specialists have arisen in the fields of Search Engine Optimization (SEO) and Search Engine Marketing (SEM), which focus on increasing the visibility of your site to the search engines. The efficacy of these methods is coming into question. The current trend is Persuasive Architecture (PA), designing a site to get the visitors to take action, such as converting them to sales. How do you know whom to engage?

Start by talking to businesses you know that are successfully marketing their products or services on the internet. These might be in your industry or a complementary industry. Ask if they will introduce you to their design firm.

Interviewing the Vendor

Picking the vendor to design your website is much like picking any other IT service provider. It isn't a quick 30 minute interview. Expect to spend several hours in the process of vetting potential vendors. You will want to cover the following topics:

- Experience. How long have they been doing SEO, SEM, PA and internet marketing? Ask for three or four client references. Be sure to ask for one former client whose project did not go

as well as they would have liked. The reason for asking for the one failure is to better understand how the design firm works. Each firm has a different strategy or process and not all strategies work in all cases. It is important to understand why a particular firm might not work for you.

- Communication skills, both listening and speaking. Do you feel like they understand your expectations for the website and its place in your business? Do they understand your marketing plan and your target market? Or are they cookie cutter designers? Just one more web site on the assembly line.

- Focus. Is their primary focus on you and your business strategy and marketing plan, or on artistry and flashy technology? You can tell by their questions and what they emphasize about the sites they show you.

- Connections. Are they a full service vendor? Can they handle the website branding, graphic design and marketing aspects as well? With whom do they partner to provide full service? Are they willing to be responsible for managing those relationships?
    - On a side note, you should register your own domain names and pay for them under your company name.

  - Also, if you are dealing with a small web designer make certain you set up and own the hosting account in your company name. This protects you in case of the financial failure of the primary vendor.

- Technology. Are they up on the latest technology trends and methods, such as Persuasive Architecture design and search engine friendly shopping carts and affiliate programs? Feel free to ask them directly. Some firms are loyal to a software provider, pushing the solutions that pay the highest commissions without regard for the needs of your business.

- Security. If you are planning to accept payment across the internet, is the vendor who is hosting the site 100% PCI-DSS compliant? Does the vendor use open source software?[1] Open source software, while popular among developers, can be vulnerable to hackers. How are they going to protect you and your customers?

- Transition planning. What is going to happen to you if their business should fold, particularly if they are hosting your website or have developed proprietary software or strategies for you? How are they going to make it easy for you to transition to another provider? If they brush off your concerns, walk away.

- Disaster preparedness. Has the provider ever experienced a disaster (a virus, lost critical files or natural disaster)? If they are hosting your site, what are they doing to ensure your site comes back online quickly after a disaster?
  - Do they have a disaster plan and have they tested it? Or do they simply have insurance on their equipment? Ask for verification and validation of their results (a SAS 70 audit will be sufficient).

Interviewing the References

The topics are very similar once again to those you covered when interviewing the references of your IT service provider.

- Results. Ask the references what results they achieved. Are they consistently placing higher in the search engine results? Do they have more of the right kind of visitors? Are sales conversions increasing? Do they feel they received value in exchange for their dollars?

- Relationship. Is the vendor responsive to changes in the business? Are they proactively keeping up with trends in your industry?

- Communication. How open are the channels of communication? Is the vendor receptive to your ideas and thoughts? How do they respond to problems with their service?

- Connections. If the vendor uses others to fill gaps in their service, how satisfied are you with the other vendors?

- Transition Plan. What has the vendor done to make it easier to transition to a new vendor? Do you have a copy of all of the applications, processes and associated documentation? Or are they keeping vital information from you?

During your conversation, be sure to notice the strategy employed with each customer you interview. Pay attention to the results they achieved working with the vendor.

Vendors to Avoid

Be sure you are comfortable with the vendor. Do you feel you can have a long-term working relationship? SEO, SEM, Persuasive Architecture and search engine friendly strategies are not overnight solutions. They all require long-term efforts to analyze results and refine the strategy. If you're not comfortable, don't enter into an agreement.

Avoid vendors that guarantee top search engine rankings. They can't control where your website places in the search results. If they say they will guarantee your placement, run.

Avoid vendors who charge by the hour, expect full payment up front or who don't provide project estimates. Expect to pay fifty percent (50%) upfront to start your project. Payment of the remaining fifty percent (50%) can be negotiated.

Last Word

Application design and website design projects have a tendency to take more time and more money than planned.

Developing a website or any other software application is like building a house. The architect has the blueprints and brings in the general contractor. The general contractor lays the foundation, builds the framework and installs the roof. As far as he knows, the house design is set. No changes in sight. Then he brings in the sub-contractors to install the wiring, pipes, cabinetry, flooring and other finishes.

Then you, the owner of the house, see something neat at the Home and Garden Show that you want to incorporate into your new house. If it is truly an inconsequential change, the architect or contractors will approve it. However, if it requires them to tear up the foundation or rewire the house, that's another story. Moving a toilet after the concrete has set is not inconsequential! You're going to pay for the effort and materials to incorporate your new idea. If you continue to tweak the design, eventually you will pay as much for changes as you planned for the original house, AND you still won't have a new house.

In IT, we call this Scope Creep, and it is completely within your ability to control. Before you engage the vendor, agree on the scope of the project. Define exactly what the vendor will do, the results they will produce and milestones for payment. If a change becomes necessary, whether the vendor realizes it or you do, you need to agree to the change and establish the cost of the change.

If you have a bright idea you want to incorporate, the vendor needs to apprise you of the cost and time delay associated with the change. Each change order should be documented (who, what, when and why) to avoid disputes and to provide a paper trail for unplanned expenditures.

Expect some scope creep. Even the best vendors who do a fantastic job of assessing your requirements miss something. They aren't omniscient. They can't read your mind. They can't think of everything. They can't predict the future. Trends within your market may necessitate a change in your strategy. Most vendors include in their project estimates an up charge of 10% to 15% to account for some scope creep. They expect overruns and have a good idea of how much it will cost as a percentage of the project's overall cost.

Remember, most scope creep is created by the client, not the vendor. Vendors are happy to satisfy your change requests, but it will come at a cost. How deep are your pockets?

1. Payment Card Industry - Data Security Standards. For more information go to http://www.pcisecuritystandards.org

I.T. Doesn't Happen To ME... by Jaime Buckley

Chapter 9

My Last Upgrade Was 1999

I walked into the office of a non-profit to assist with a financial package upgrade. Their last update was 10 years ago, in 1999. A law office upgrades their financial and time tracking software every two or three years. A manufacturing firm upgrades their accounting software every couple of years. This is typical. Most small businesses function quite well without the latest, greatest technology.

Small businesses and non-profits tend to be very stable in terms of their technology architecture. Once it is up and running, they don't feel pressed to upgrade their hardware or applications as long as their needs are being met and their financial software is legally compliant. That's good. It keeps technology expenditures low.

Planned obsolescence is a dirty little secret of the IT industry. It enables vendors to respond to their customer's needs and, more importantly, remain profitable. Software vendors release new versions of their products every year or two. When a new version is released, they automatically stop supporting the oldest versions of software. In general, vendors will only support the two most recent versions of a product. If you have a problem with your older software, you're on your own.

The best arbiters of whether or when you should upgrade are your business strategy and your disaster recovery plan. Does your business need the most-recent upgrade to remain competitive? If not, you can probably wait a while. On the other hand, if a software application is a critical component of a vital business process, you should upgrade the application regularly to minimize disaster recovery efforts. If you don't plan to stay fairly current, be sure you keep the application installation discs somewhere safe offsite.

Small businesses tend to keep the same operating system until they purchase new hardware. Generally, if your operating system is more than three versions old, it is probably not supported. If you have problems with applications or the operating system itself, the vendors will insist that you upgrade your operating system to resolve the problem. At that point you have no other option.

You can introduce a new computer and operating system into a network of older computers. The new computer will function fine in your network. You can install the older applications on the new computer, provided they are compatible and you remove them from the decommissioned one to avoid software piracy issues. If your new computer comes with newer versions of application software, you can set the default options, so the application files are saved in the older versions for compatibility throughout the office.

When addressing application upgrades the same thoughts apply. You will need to know the vendor's support policies. Should you wait ten years between upgrades? I don't recommend it. Quite often the upgrade path is neither direct nor easy over that length of time. It could prove costly. If the application is critical to your operation, also consider the issues mentioned in disaster recovery.

The best advice on upgrades: let your business strategy and your vendor's support policy guide you. You don't need to be on technology's cutting edge.

However, you should not stay so far back that an upgrade results in significant expense (e.g. new computers as well as applications) or significant delays in disaster recovery.

GRANTLAND®

HOW DID YOU GET CAUGHT?

THEY WENT THROUGH ALL OF MY FILES, ALL OF MY RECORDS, ALL OF MY TRANSACTIONS --

THEY CHECKED AND DOUBLE CHECKED EVERYTHING!

CAN YOU BELIEVE THE LEVEL OF MISTRUST SOME PEOPLE HAVE?

Chapter 10

Software Piracy

Of all the IT mines you can encounter, software piracy is probably the most pernicious. Seemingly innocent actions can cause great harm to a business.

According to the 2007 Global Software Piracy Study, sponsored by the BSA and IDC, 38% of the world's software is pirated resulting in $47.8 billion in losses to the vendors.[1]

If you are caught with pirated software, it can cost you dearly with fines and imprisonment. The penalties could be severe enough to sink your business, and the humiliation could ruin your reputation.

In addition, pirated software is often the vehicle for malicious software that can destroy your data and take down your network.

Software piracy is copyright infringement. All software vendors are afforded certain protections under

U.S. copyright law for their software. The vendors in the End User License Agreement (EULA) tend to define it as the unauthorized copy and distribution of their work product. Their rights are subject to limitations of the fair use doctrine which is reproduced below.

> Notwithstanding the provisions of sections 17 U.S.C. § 106 and 17 U.S.C. § 106A, the fair use of a copyrighted work, including such use by reproduction in copies or phonorecords[sic] or by any other means specified by that section, for purposes such as criticism, comment, news reporting, teaching (including multiple copies for classroom use), scholarship, or research, is not an infringement of copyright. In determining whether the use made of a work in any particular case is a fair use the factors to be considered shall include:
>
> 1. the purpose and character of the use, including whether such use is of a commercial nature or is for nonprofit educational purposes;
>
> 2. the nature of the copyrighted work;

> 3. the amount and substantiality of the portion used in relation to the copyrighted work as a whole; and
>
> 4. the effect of the use upon the potential market for or value of the copyrighted work.
>
> The fact that a work is unpublished shall not itself bar a finding of fair use if such finding is made upon consideration of all the above factors.[2]

I presume that you are of good moral character and would not knowingly pirate software. Unfortunately, it can happen without your knowledge. Consider these two cases:

Case 1: An employee received a call from a friend at another company. The friend needed help recovering a mainframe system; it was down and not coming back up. His employer had the software his friend's company had not yet purchased or licensed. He reasoned that his company would deny any request for assistance to help recover the mainframe. Time was of the essence as the machine needed to be up by Monday morning. He took the software to his friend and they recovered the system.

Case 2: A disgruntled employee left his employer and took proprietary company software. He then installed the software at his new company.

Did copyright infringement occur? Yes, in both cases.

In the first case, the software vendor lost a sale. Due to one employee's actions, its potential market was affected. Furthermore, if the problem remained undetected the vendor would have been denied future licensing fees. With a bootleg copy of the software floating around, the possibility for a repeat incident was also high.

As soon as the problem was discovered, the IT Department heads from both companies and their legal teams sat down with the vendor and worked out an equitable solution. The two employees involved were sanctioned. While their intentions were good, their actions could have caused major problems for both companies.

In the second case, the intent was clearly malicious. The company was not denied income from the sale of the software, but their competitive advantage was compromised. When the company called asking for help in making an update to the software, the breach was discovered. They were told to remove the software. They complied.

While these cases are not that unusual, according to the BSA, most piracy occurs simply through the over-installation or overuse of software. Over-installation occurs when you install software on more computers than the number of purchased licenses you hold. Overuse occurs in a client-server environment where a centralized copy of software is in use by more users than licenses held.

During the 1990's many large and reputable companies were caught up in the explosion of desktop computers and servers. The desktop PC was intended to free the user and departments from the perceived bureaucratic practices of the mainframe. Before the frenzy subsided, many companies realized they had an unintentional piracy problem by purchasing fewer software licenses than they were using.

Failure to coordinate ordering and installing software resulted in the over purchasing and sometimes under purchasing of software licenses. It cost them millions to come into compliance and implement centralized purchasing and installation practices.

As you can see, piracy can happen right under your nose without you being aware. Some employees with good intentions can unknowingly infringe on a copyright through trading or downloading software from seemingly reputable dealers.

Some employees, thinking they are helping you control expenses, may bring in software from home or another job. Others will be malicious and take your software and license number home, to another employer, or resell it. It doesn't really matter how it happens. As the business owner, you bear the responsibility, so you must ensure that it doesn't.

Not all copying or reuse of software constitutes a copyright infringement. You are allowed to create a copy of the software for backup purposes such as disaster preparation or to repair a failed computer.

Software reuse is also allowed in certain cases. You may copy software from one machine to another when you purchase a new machine, provided that you remove the software from the old machine or the old version of the software is no longer on the market. Most software packages will require you to redistribute the licenses if your use exceeds the number purchased.

If you copy software to a new machine and leave the old one in service, you have effectively denied the vendor income from the sale of the software thereby adversely impacting his market. You are guilty of software piracy. Do the right thing. Buy a new copy or license for the software.

Generally, if you have purchased a software upgrade, you can give away the old version. The market for the software has not been substantially diminished by the gift provided the older software is no longer on the market. You paid for the first version as well as the succeeding versions and in the installation, removed the prior version.

It is your responsibility to protect your company from copyright infringment.

- As with your security policy, define and clearly communicate your software purchase and installation policy to your employees. Do not create a policy that is so bureaucratic that your employees are looking for ways to circumvent it in order to do their job.
- Lock down company computers to prevent unauthorized software installation.
- Centralize software purchasing. Create a software asset inventory to track the number of licenses or programs purchased and how they are distributed among your users.
- Store the software discs and licenses in a secure place to prevent inappropriate usage. The discs, licenses and invoices are proof of purchase.

- Periodically perform a company-wide software audit. For a small company, an annual audit is generally sufficient. The audit simply compares the software on each computer to the software asset inventory. For a small environment, a hand audit is fairly easy. The BSA provides links to several free audit tools (http://www.BSA.org) in the Anti-Piracy section of their website.
- Ask your software developer to include anti-piracy measures in proprietary software.

If your company is accused of piracy, the authorities don't usually come looking for your employees. They look for you, the owner. Ignorance of what has happened inside your company is not a mitigating factor. Implementing and enforcing appropriate software security policy is the most effective protection.

Please note: I am not an attorney offering a legal opinion. I am not responsible for damages you incur as a result of your interpretation of copyright law or what I write. If you have specific questions about software piracy, please consult an attorney.

1. For more information on the study, visit http://www.BSA.org
2. US Code: Title 17,107. Limitations on Exclusive Rights: Fair Use

GRANTLAND®
....SO WHEN I ASKED HIM WHAT RISK MANAGEMENT IS IN A NUTSHELL,
HE SAID IT'S LIKE MAKING SURE YOU HAVE A SPARE TIRE WHENEVER YOU GO ANYWHERE.
SO HAVE YOU TRIED IT YET?
OH YES,
IT'S NOT NEARLY AS COMPLEX AS HE FIRST MADE IT SOUND.
Copyright Grantland Enterprises. All rights reserved
www.grantland.net

Chapter 11

Clearing the Minefield

Now that you have exposed the mines in your IT environment, you are ready to defuse them to protect your business.

Your IT Manual

Your IT Manual will contain documentation about the business IT environment and work processes.

The support CD for this book contains a file you can use to create your company's IT Manual. The file is formatted for printing on 8.5" x 11" paper with a 3-hole punch so that you can keep it in a 3-ring binder. You have the right to reprint as many pages as you might need for your own use. If you wish to supply this CD to a client or another small business owner, please direct them to www.ITMinefield.com to purchase it legally.[1]

The Table of Contents is formatted to print on an Avery Table of Contents with 15 tabs. Feel free to re-order the sections or add sections to suit your purposes. Many sections contain sample forms for your use.

The sections of your IT Manual include:

- Hardware Asset Inventory
- Software Asset Inventory
- Network Management
- Problem/Service Log
- Disaster Planning
- Vendor Contact Information
- Website
- Work Process Documentation
- New Employee
- Policies
- Priorities & Action Plans

Re-evaluate

Review your answers to the IT Mine Detector™. Which areas did you rate as important or very important (a value of 2 or 3) and yet you rated your confidence as low (a value of 1 or 2)? Considering what you now know, do you need to re-evaluate your responses? Did the importance of an area change? Did your level of confidence change? Which of these areas now seem important to your business?

Cost Management

Cost management is a constant theme. As in many businesses, the cost to fix something is greater than the cost to do it right the first time. Consider these questions:

- Do you have a strategic plan?
- Is your IT documentation in order?
- Do you have a security policy?
- Do you have a process for managing change in your environment?
- Do you have a process for developing or changing applications?

- Do you have a disaster recovery or business continuity plan?

Many of these areas overlap. For example, security touches network, disaster planning and software legal compliance. Change management involves the operating systems, business applications and website development. Documentation crosses several areas.

Which of these areas seems most important now?

Documentation

Documenting the environment is necessary and not that difficult, but it can be time consuming. Without the documentation, you have no idea how to rebuild when needed.

Documentation might not be a task best handled by you personally. An IT professional can help if needed. Present him with your IT Manual and ask for a cost estimate to complete the documentation you require. Be sure to include the hardware and software asset inventories and the network section.

Disaster Planning

Disaster planning is about risk mitigation, doing what you can to minimize or eliminate delays in recovery. It is not possible to create a 100% solution that would make disaster recovery completely automatic, invisible and painless, but you can make it easier.

Start by asking these questions:

1. What are our business critical processes?

2. What are the technology components on which these processes depend?

3. How long can the IT environment be unavailable before the business is adversely impacted?

4. What can happen to cause a mission-critical process to quit working?

   a. File deletion
   b. Virus or hack attack
   c. Disgruntled employee
   d. Hard drive crash
   e. Server or other hardware failure
   f. Hosted application failure
   g. Network/internet access failure
   h. Power failure
   i. Facility destruction

When considering processes, are critical components beyond your control? For example, application hosting: If your website or other critical application is hosted by another company, can they recover from a disaster, whether it be a lost file, server or facility? Can they verify their plan has been tested successfully?

Looking at your answers, "ask" what can you do to decrease delays during recovery? What can be done to mitigate risks?

The risk associated with viruses and malware can be reduced by several different actions: establish an appropriate use policy, lock down the desktop to prevent unauthorized software installs, block websites that are not related to your business, install hardware firewalls, etc.

While file backup does not eliminate a risk, it does reduce delays in recovery. Appropriate HR and IT policies can prevent a disgruntled employee from doing damage by quickly changing access permissions to the facility and files.

Most of the preventive actions you can take are straightforward and rooted in common sense. Don't become overwhelmed by trying to address all of the possible disasters at once. Start with the ones you think present the highest risk, and are most likely to occur. If you need help, contact an experienced IT service provider.

Once you have a plan, work with your provider to test it. An untested plan is worthless.

The disaster planning section of your IT Manual has forms for helping you work through the process.

Security Policy

Security policy is a very broad subject that covers a number of different areas. Depending on the size of your company, you will need policies for:

- Acceptable Use: This is the broadest of all of the policies covering everything from personal use of computers and the network, copyright infringement, to unauthorized software installation and much more.
- Passwords: This policy defines the standards for passwords in your environment including password length, composition and frequency of change.
- Data Backup: The backup policy defines mission-critical, important and archival files and the frequency of backup. This policy anchors your Disaster Planning.
- Software Purchasing/Installation: The SPI defines the process for bringing in new software

to help you manage the proliferation of software in your environment. It defines who is responsible for ordering, receiving, and completing the install.

- Network Access: This policy covers who may access the network, the standards for access, the use of wireless devices and wireless access.

- Network Security: covers anti-virus software use and maintenance, confidential data access, security of network appliances (firewalls, switches, wireless access points). This is a very broad and very technical policy designed to protect your business from all of the evils that can assault it. Even if you are not ready to implement this policy, having it can help you identify potential vulnerabilities.

These definitions are very broad. Some service providers will divide them up differently, combining some and separating out others.

If you don't want to write a security policy you can hire a company that specializes in creating security policy for the small to mid-sized business. Some of them automate the process by asking you specific questions that will generate a policy tailor-made for your business.

Don't use security policy to create a bureaucracy that hinders your employees' ability to do their jobs. Just put enough in place to protect your business.

Website

If your website is being hosted by a service provider then consider:

- Are the domain names registered in the name of your company?
- Is the hosting account registered in the name of your company?
- Do you have a copy of all of the documentation, pages and processes for the site?
- Can the hosting service verify their disaster recovery plan actually works?

If you are hosting the site:

- Have you worked with the website designer to develop a disaster recovery plan?
- Are the files associated with the website part of your backup plan?

Maintenance

If you permit automatic updates to occur, your systems will stay current on maintenance. If not, you need to work with your service provider to schedule routine maintenance. Routine maintenance often includes security patches that are necessary to protect your environment. Completely ignoring it will leave your systems vulnerable.

Legal Compliance

There are many steps you can take to minimize your exposure to software piracy issues such as:

- Complete a software asset inventory and periodically conduct an audit of your computers. The IT Manual contains a form you can use to manage your inventory.
- Lock down the desktop or laptop to prohibit new software installs.
- Prevent access to questionable websites.
- Centralize the purchase and installation of software.
- Minimize red tape that makes it difficult for employees to get the tools they need.

It is easy to keep the inventory up-to-date when it is a step in your company's purchasing or installation process. Perform a software audit once a year. If you should discover discrepancies between the inventory and software installed, then:

- Take appropriate action to become compliant. Remove all unauthorized or unlicensed software or pay for it.
- Discipline the offending parties. If you are the offender, then remember, what goes around comes around. If you are dishonest, then you will eventually find yourself surrounded by dishonest people.

Review your policy annually to ensure it is still appropriate to your business environment.

Next Steps

Considering what you now know, where do you need to focus first to protect your business? You may feel that they are all important, but for now, just pick one and create an action plan.

An action plan documents a sequence of events necessary to achieve a larger objective. It starts with a specific goal or stated objective. "Preventing disasters" is not a goal. It states an intention.

An appropriate goal would be: Create and test a disaster recovery plan by December 31, 2010. Goals need to be SMART: Specific, Measurable, Attainable, Realistic and Time bound.

For each goal, list the steps that are required to complete it. Each step states a specific action, declares who is responsible, when it is to be done and identifies any obstacles or barriers to completion. Some steps may have several sub-steps.

Using the previous goal as an example, you might have some of the following steps or smaller, subordinate goals:

1. Identify critical processes and components by January 31, 2010.
2. Develop storage and backup policy by February 28, 2010.
3. Engage a vendor to provide offsite backup by March 31, 2010.
4. Audit the IT environment to discover any potential vulnerability by July 31, 2010.
5. And so on.

See the Priorities & Action Plans section of your IT Manual for a sample form you can use for your planning purposes.

After you have completed the work on the first priority, move on to the second one. You don't need to accomplish everything in one month or even one year. Do what you can to defuse the mines one by one starting with the one you feel is most dangerous.

Whatever you do, don't:

1. If you wish to purchase the support CD mentioned within please contact us at www.itminefield.com

Conclusion

Looking back, looking ahead

I have found it interesting to watch the world of Information Technology evolve over the last 20-plus years (and I'm still a fairly young person). I started my career in the mainframe arena where part of my job was to develop processes and practices designed to maintain high reliability and availability. Due to the centralized nature of computing, we had to be disciplined in our practices. Crashes within a multi-million dollar mainframe at a Fortune 50 company were not smiled upon.

The advent of the wild and wooly world of desktop computing, servers and networks changed the landscape of computing and support. Often it meant these systems were implemented without much thought to prior practices and why they were necessary.

"Flexibility!", "Power to the user!" and "Death to the mainframe!" were the battle cries of the day. Predictably, businesses encountered a few mines resulting in major crashes and data loss. It wasn't long before the desktop and server world realized the need for the discipline and rigor of mainframe practices.

Now we are in the era of the "small business", where small describes the size of the computing footprint, not company revenues. The path through the minefield has been cleared, yet the lessons learned from the previous two eras have been slow to reach the small office.

As I said earlier, I am surprised at the number of small businesses I visit that don't even have basic processes and documentation in place. IT support people often treat the small office as if it needs less discipline because it is small. Yet nothing could be further from the truth. As our economy is increasingly driven by the smaller enterprise, the ramifications of IT failure are far-reaching. The small business must begin to employ the discipline of previous eras to ensure its wellbeing and survival. Information theft or loss, in whatever form, can kill a small business.

You may choose to outsource many IT related tasks to a service provider. That's fine. Outsourcing is probably a more reasonable and strategic use of resources rather than hiring IT support in-house. However, you cannot abdicate your responsibility by tossing it to the service provider with a "Here, you take care of this. I don't understand it and I don't need to," attitude. Your intimate knowledge about your business is invaluable to the service provider if he is to be effective. As small business owners, it is your responsibility to communicate your needs and hold your service providers to a standard of discipline that protects your enterprise.

Wishing you safe passage.

About the Author

Leslie Knight is the founder and General Manager of Knight Performance Management, a business focused on improving performance and decreasing costs through effective strategy execution and team development.

Leslie consults with corporations to integrate their IT strategy and expenditures with their overall business strategy. She applies over 20 years of experience developing, planning and managing Information Technology at larger corporations to her work with entrepreneurs.

Her advice has saved Fortune 500 Companies millions of dollars. Leslie has been recognized by past clients for an uncanny ability to:

- Quickly size up a situation
- Formulate a plan of action based on available information
- Put together a team
- Lead the team to execute the plan to successful conclusion

Customers value her unbiased point of view and concern for the wellbeing of their business.